Dedicated to our inspiring friend
ROBERT E. SNODGRASS
who shared his creative nature.

Special thanks to the David and Lucile Packard Foundation.

Copyright © 2000 by Dawn Navarro
All rights reserved, Published by Sea Challengers in association with Manta Publications.

Dan Gotshall, Editorial Director
Dawn Navarro, Art Director-Book Designer-Illustrator
Wallace 'J' Nichols, Scientific Advisor-Writer
Roxane Buck-Ezcurra, Scientific Editor-Story Editor
Dominique Navarro, Book Editor
Robert Cuoto, Electronic Production Consultant
Wendy Ray, Book Production Associate
Ann Gotshall, Advertising-Publicist

No part of this publication may be reproduced in whole or in part, or stored in a retrieval system or transmitted in any form or by any means electronic, mechanical, photocopying, recording or otherwise without written permission of the publishers. For information regarding permission write Manta Publications, 30069 Harvester Road, Malibu, CA 90265.
Printed in China through Global Interprint, Inc., Santa Rosa, CA.

Library of Congress Cataloging-in-Publication Data
Navarro, Dawn.
Chelonia: return of the sea turtle
p. cm.
Hardbound: ISBN: 0-930118-31-6 / Soft-cover: ISBN: 0-930118-41-3
1. Green turtle--Mexico--Baja California (Peninsula)--Biography--Juvenile Literature.
2. Wildlife rescue--Mexico---Baja California (Peninsula)--Biography--Juvenile Literature.
[1.Green turtle. 2. Turtles. 3. Wildlife rescue.] I. Title.
QL666.C536 N28 2001
597.92--dc21
00-066172

Watch for the sea glossaries.

BASED ON A TRUE STORY

CHELONIA
Return of the Sea Turtle

Dawn E. Navarro
Robert E. Snodgrass
Wallace J. Nichols

MANTA
PUBLICATIONS

SEA CHALLENGERS

It was midday in August 1971. An El Niño tropical storm was brewing to the south.

Nina Delmar and her father navigated their sturdy yacht on its homeward cruise off Mexico's coast. Wind-swept clouds darkened the sky above. Beneath, the deep waters churned and rolled.

They watched the sea with steady eyes as they faced the most difficult passage of their journey.

SEA LOG

Five hundred miles south of the U.S. border, in Baja California, Mexico, the sea is big and blue. Dolphins swim and play here. Sea turtles from Michoacán nesting beaches come to feed on crabs and sea grass. Giant whales feed on fish and krill. The cruise north off the Baja coast is infamous and appropriately named "the uphill passage" by seafarers. The name describes the climb boats must make to the top of each wave coming out of the north.

Bottlenose Dolphin
Tursiops truncatus

White-sided Dolphin
Lagenorhynchus obliquidens

Common Dolphin
Delphinus delphis

Nina's father felt the fury of the sea. "We can't beat this storm before dark," he shouted. "We'll have to ride it through the night."

"The dolphins are gone, Papa," Nina cried out over the howling wind.

An albatross grazed its wing on the crest of a wave rising before them. Rain clouds blocked the last rays of sunlight and the sea turned angry gray.

That night Nina hid in her tiny bunk below deck. She wished that she could dive deep below the waves like a dolphin, or fly away like a seagull.

SEA LOG

During storms, dolphins can dive beneath the waves to calm water. They return to the surface only to catch a breath of air. Birds often glide just above the waves using the rising crests for lift. Both dolphins and birds are adapted to survive in rough weather, but some storms are difficult in the open sea, especially for the young. Only the strong and lucky survive.

Storm Petrel
Oceanodroma leucorhoa

Albatross
Diomedea albatrus

Seagull
Larus occidentalis

Through the night the storm's strong current swept a young sea turtle far out of its normal range. The little animal struggled hour after hour in the dark and chilling sea.

With the light of dawn the storm had passed. The weak and weary turtle bobbed on the surface, barely able to lift its head for a breath of air.

SEA LOG

Tunas, sharks and other fish of the open ocean can be found hanging out underneath floating objects. In their world there are few hiding places; these objects may offer them protection. They also follow warm water currents which extend especially far to the north during the warmer El Niño years.

By mid morning the sea had calmed. But the hapless little turtle still rolled in the smooth waves.

A hungry tiger shark, scavenging after the storm, swallowed anything and everything he found. He spotted the sea turtle and moved towards his prey. However, something disturbed his hunt. The shark quickly circled deeper as a rumbling noise grew louder.

Just then, the Delmar's yacht crested a wave.

SEA LOG

After a storm, drifting debris, seaweed rafts and unlucky or stranded sea creatures accumulate along the edges of currents. The ocean's scavengers follow these "green lines" looking for a meal. In this way, sharks and seagulls clean up after a storm.

Surf Grass
Phyllospadix spp.

Giant Kelp
Macrocystis pyrifera

From the bow of the boat Nina kept watch on the sea, entranced by the excitement in the water. Tunas darted among the leaping dolphins. Whales splashed and sprayed.

Ahead she spotted an unusual shape. It looked like a small turtle to her. It was!

"Papa, look! A sea turtle," she cried. "It's looks hurt... can we help it?"

Papa turned the boat toward the floating animal. "Take the helm, bring her about, full stop!" he shouted.

Nina took control of the boat as her father rushed to the stern, stripping off his shirt and leaping into the cold water.

Papa grasped the helpless turtle with both hands. Nina threw him a life ring and helped pull her father and the rescued sea turtle to the boat.

Papa shivered as he climbed aboard. The tiger shark passed below unnoticed.

Nina wrapped her shivering father and the cold turtle in warm, dry blankets.

"What can we do?" Nina asked as she held the fragile bundle. The turtle was so weak it could hardly move. Its eyes were barely open.

"It might not survive, Nina. Let's put the turtle in this wooden box and let it rest," her father said gently.

Nina watched over the turtle for hours. She kept its eyes moist using an eyedropper.

It was late afternoon when they neared the California coast.

SEA LOG

All around the world, wildlife rescue centers work to save injured or stranded animals. It is everyone's responsibility to rescue animals in need of help. Whenever a tagged animal is encountered, the local stranding or rescue center should be contacted. Trained veterinarians work with the animals to rehabilitate them and return them to the wild.

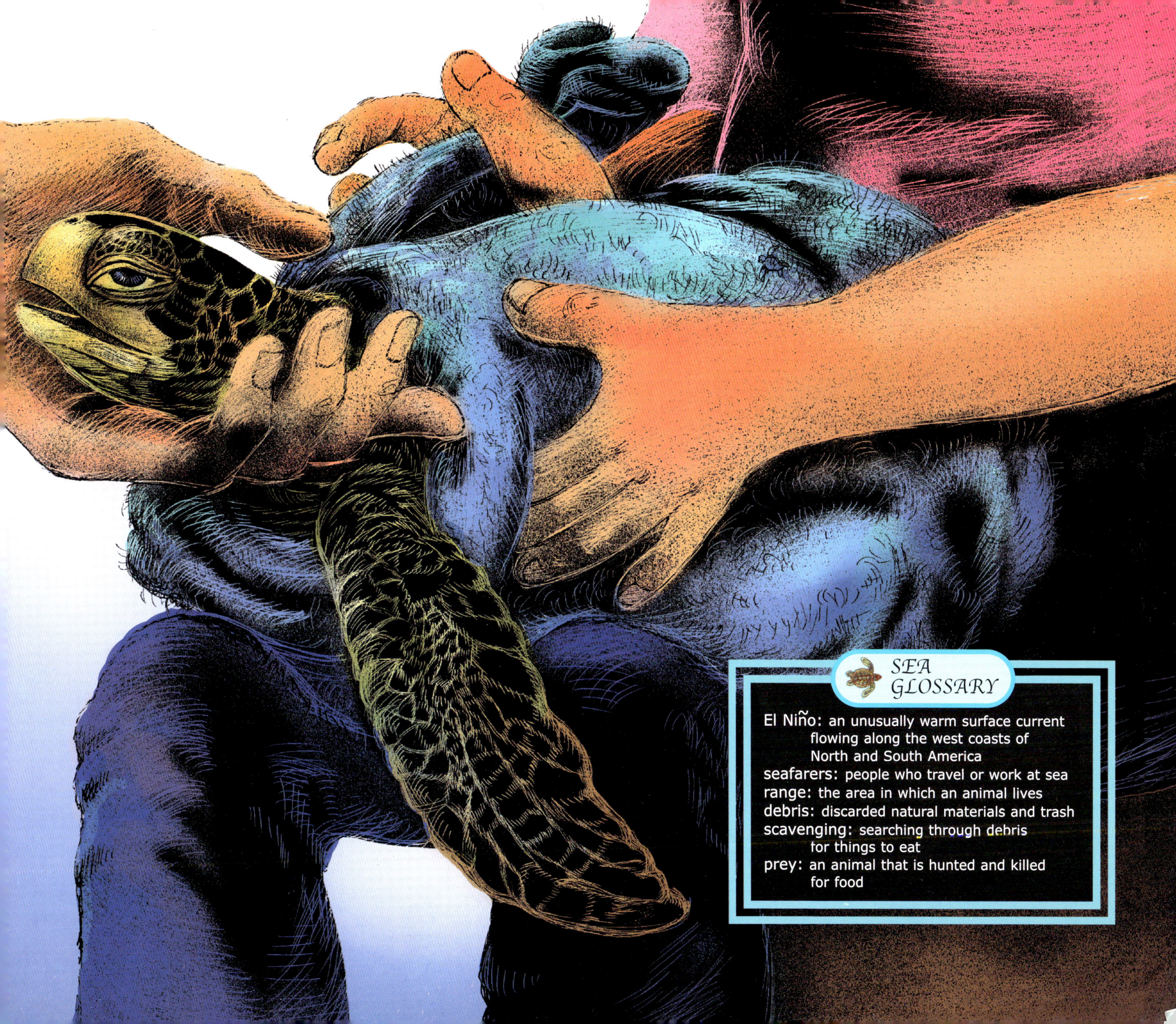

SEA GLOSSARY

El Niño: an unusually warm surface current flowing along the west coasts of North and South America
seafarers: people who travel or work at sea
range: the area in which an animal lives
debris: discarded natural materials and trash
scavenging: searching through debris for things to eat
prey: an animal that is hunted and killed for food

By late evening the Delmars' yacht entered San Diego Bay. They docked at the U.S. Customs office.

"We have something to show you," declared Mr. Delmar.

"After the storm we rescued this baby sea turtle," Nina shared.

The customs officer looked at the small, limp turtle in the wooden box. "It's a green turtle, 'Chelonia mydas.' What a shame. There aren't many left these days. It doesn't look like this one will survive," he commented as he wrote on his clipboard.

"Chelonia," Nina whispered. "Don't worry, I'll take care of you."

When Nina and her father arrived home they were exhausted. But Nina filled their bathtub with warm water. This was a safe place for the sea turtle to spend the night.

Nina instantly fell asleep in her own comfortable bed. All night she dreamed of the storm, the dolphins and the seabirds. She imagined the sea rolling around her as if she were still on the boat.

SEA LOG

Chelonia mydas (pronounced ki-'lō-nē-a mī-das) is the scientific name for the green sea turtle. The Pacific green sea turtle is found in coastal waters, bays, lagoons, estuaries and the open ocean. They migrate thousands of miles with the currents along the Pacific Ocean coasts from nesting to feeding grounds.

The next morning Chelonia was revived by the warm water and rest. She swam and splashed in the bathtub. Her eyes were wide open with curiosity about her strange new environment. She was hungry and tried to bite a rubber ducky.

Nina spoke to the sea turtle as she took her into the backyard. "You need more room and salt water and food. You're going to like our pond, Chelonia."

SEA LOG

Sea turtles swim using their flippers to pull themselves through the water. Both flippers are used at the same time, moving up and down, much like a bird flying. The rear flippers are used as rudders to steer and to move backward. Turtles can swim fast and dash away quickly to escape danger and predators or to chase their prey. During migrations, green turtles may swim more than 20 miles each day.

By-the-Wind Sailor Jellyfish
Velella velella

The Delmar's backyard saltwater pond was home to a small garibaldi, a few reef fish and three young green moray eels. It was a comfortable new world for the small sea turtle who swam around exploring the rocks and resting on the sandy bottom.

Nina fed the turtle bits of fish and handfuls of sea grass she had collected in the ocean. In the months that followed, the little sea turtle grew bigger and bigger and bigger.

SEA LOG

In the wild, green sea turtles eat mostly algae and sea grass. They occasionally eat sea pens, sponges and jellyfish. In captivity, sea turtles are fed fish and squid and may grow two or three times faster than in the wild.

Sea Pen
Ptilosarcus gurneyi

Red Algae
Gracilaria sjoestedtii

All of the sea animals were well fed and grew quickly in the backyard pond. Chelonia could barely turn around she had grown so fast.

In just two years Nina's small turtle had grown into a large wild animal.

"Chelonia, how big do green sea turtles grow?" Nina wondered, "And what are we to do about you? You need a bigger home."

5 years

20 years old

Hatchling

SEA LOG

When green turtle hatchlings are born, they are smaller than the palm of your hand. It can take a green turtle 5 years to reach the size of a small platter. It may take them more than 20 years to reach maturity (approximately 30 inches). Green turtles can grow to be more than a yard long (300 pounds) and may live for 100 years.

Nina studied a book about sea turtles and learned that there are only seven species of sea turtle in the world. All are endangered. She realized Chelonia needed to be returned to the sea. Nina called her Aquarium for help.

"Hello, my name is Nina and I have a sea turtle in my backyard."

Loggerhead Sea Turtle
Caretta caretta

Olive Ridley Sea Turtle
Lepidochelys olivacea

Kemps Ridley Sea Turtle
Lepidochelys kempii

Australian Flatback
Natator depressus

The Aquarium's marine biologist told Nina that she had done a great thing by saving Chelonia's life. He also explained to her that the 1973 Endangered Species Act now protects all sea turtles. "The next time you find an injured animal, please contact us right away. Chelonia is a female turtle. Returning her to the ocean will help her species survive," he said. "We'll watch her behavior, check her health and tag her before we release her into the ocean."

A few weeks later Nina was invited to help release the healthy, strong green turtle. They sailed out on a small boat into the Pacific Ocean. An underwater photographer documented Chelonia's release as the turtle quickly disappeared into the blue sea.
"Vaya con dios, Chelonia."
Nina waved goodbye.

Chelonia instinctively remembered everything about being a wild turtle. She swam south toward the warmer waters of Mexico eating jellyfish carried by the currents. While following the path of her ancestors, she avoided fishing nets that often drown sea turtles. Chelonia headed for the protected beach where she was born in Michoacán, Mexico. Green turtles have nested there for millions of years.

SEA LOG

Sea turtles thrive on jellyfish. But they often can't see the difference between drifting plastic bags and their jellyfish prey. When a sea turtle swallows plastic debris it will often cause the turtle to suffocate and die.

Chelonia traveled for many months and swam thousands of miles in the ocean. She finally arrived at the beach where she was born. Hundreds of sea turtles gathered in the shallow waters. There were female and male turtles, brothers and sisters, cousin and nephew turtles. There were even aunts and uncles. Some of the turtles were 100 years old. All of the sea turtles came for the same reason—to mate and lay eggs.

SEA LOG

Adult male and female sea turtles can easily be distinguished. The males have long tails and big, curved claws. However, when the turtles are young it is difficult to tell the males from the females.

Female **Male**

One moonlit night, when the time was right, Chelonia moved through the waves and dragged herself up the sandy beach. She dug a deep pit with her rear flippers. Then she dropped one hundred perfectly round white eggs into her sandy nest. She carefully covered the eggs and returned to the sea.

Eight weeks later, on another moonlit night, Chelonia's eggs began to hatch. Baby sea turtles pulled themselves up through the sand and scrambled down the beach into the sea.

SEA LOG

Baby sea turtles are called hatchlings. Once hatched it takes the babies three days to climb through the sand out of the nest. They are able to locate the sea at night by moving toward the moonlit water. However, if there are bright lights from parking lots, roads or houses they will become confused and never find the ocean.

Many obstacles awaited as they made their journey to the ocean. Predators, such as crabs and raccoons, ambushed the hatchlings. Flocks of hungry seagulls hovered above, eyeing the turtles as an early morning meal.

Even in the water the surviving sea turtles were still not safe. They'd have to avoid the attacks of fish and sharks as well.

SEA LOG

From the hundreds of sea turtles that leave the nest only 10 percent will survive the first year of life. The dangers begin even before they hatch. Turtle eggs are savored as meals by predators including dogs, coyotes, foxes and seabirds. However, humans are the turtle's greatest predator. People dig up egg from nests and eat turtle meat despite the 1973 Endangered Species Act. The fate of the sea turtle lies in our hands. Will we give them the opportunity to live another hundred million years?

One little sea turtle would endure. And that turtle would return to this same beach someday to lay her eggs...

...just like Chelonia.

Sea turtles are endangered and disappearing from our oceans. You can help them survive by taking Nina's Sea Turtle Pledge and by supporting conservation groups working to rescue sea turtles and other marine animals.

My Sea Turtle Pledge

1. Help Our Wildlife!
   Call a professional wildlife rescue team when an animal needs rescuing.

2. Respect wild animals.
   Give them plenty of room on the beaches and in the ocean.

3. Keep our beaches clean.

4. Keep our water clean.
   No plastics and other trash in our oceans and streams.

5. Ask Mama and Papa to buy only "Turtle Safe" seafood.

6. Learn more about the wild ocean animals and tell your friends about sea turtles.

Nina Delmar

GrupoTortuguero
www.grupotortuguero.org
The Grupo Tortuguero is a network of fishermen, scientists, teachers, students, government officials and conservationists working together to save sea turtles! These sea turtle protectors monitor and study turtles, their nests and their habitats, and spread the "Save Sea Turtles!" message to their communities by teaching how important these ancient creatures are. The Grupo Tortuguero hopes that one day sea turtles will no longer be endangered and will once again swim in the waters of the Pacific Ocean by the hundreds of thousands!

Pro Peninsula
www.propeninsula.org
Pro Peninsula is an organization based in San Diego, CA dedicated to empowering individuals, communities and organizations on the Baja California peninsula to protect and preserve their unique environment.

California Academy of Sciences
www.calacademy.org
Since 1853 scientists at the California Academy of Sciences have been exploring the world to learn about plants and animals. Educators at the museum use what the scientists learn to teach visitors about nature and visitors learn how important it is to protect the natural world for future generations.

INTERNATIONAL SEA TURTLE SOCIETY

International Sea Turtle Society
www.seaturtle.org

Monterey Bay Aquarium
www.mbayaq.org

Sea Turtle Restoration Project
www.seaturtles.org

WALLACE J. NICHOLS PH.D.

Dr. Wallace J. Nichols is a scientist and defender of the ocean. As a boy he spent much of his time exploring the sea and the woods, and he fell in love with the topics of genetics, animal migration, culture and conservation. In 1998 he founded the Grupo Tortuguero, a movement dedicated to protecting the sea turtles of the Pacific Ocean and promoting responsible fishing. Currently he works with Pro Peninsula, helping to conserve nature on the Baja California peninsula, where one can find animals unique in the world such as the big horned sheep, the gray whale and five different species of sea turtle. J. also leads a program for young people dedicated to creating awareness about the threats faced by marine ecosystems all over the world, problems that can only be solved by creating an 'Ocean Revolution.' He works with several Universities and organizations that want to help preserve the beauty of nature for future generations. So great is his desire to protect the environment, he once walked 1,180 miles from Oregon all the way to the border with Mexico to bring attention to coastal and marine environmental issues! You can reach J. at j@oceanrevolution.org or at +831.426.0337.

ROBERT E. SNODGRASS

Robert Snodgrass was dedicated to the study of natural ocean history and marine biology. At the age of sixteen he became an active junior volunteer at UCSD's Scripps Institution of Oceanography in La Jolla, California. After graduation from U.C. Berkeley he returned to the Scripps Aquarium to become the Curator of Collections and Fish. An international aquarium consultant, he specialized in aquarium design and marine life support systems as well as animal husbandry and a technique of creating fish community tanks. Robert was an experienced diver who was as comfortable as a fish underwater and documented his experiences with videotape. As a naturalist he gave lectures, classes and guided natural history tours, dives and walks in the La Jolla cove for more than two decades. He also led collection expeditions and trips to the Sea of Cortez. Robert was an accomplished writer who wrote a *Los Angeles Times* syndicated newspaper column "OCEAN ENCOUNTER" for five years. He dreamed of "sharing his knowledge", experiences and stories beyond the dozens of educational books and papers produced for the aquarium. His life goal was to bring to the people- who might never "experience it firsthand"- the fragile and increasingly endangered ecosystem of the sea.

DAWN E. NAVARRO

Dawn Navarro studied illustration and design at the Art Center College of Design, Los Angeles. After graduation she worked as an art director, illustrator and designer for many companies including Walt Disney, Needham, Harper & Steers Advertising and the Los Angeles Times. As an naturalist artist, she focuses her studies on marine ecology. During a diving and fish collecting expedition to the Sea of Cortez with Robert Snodgrass, Dawn began a series of animal behavior drawings that became the illustrations for the weekly newspaper column "OCEAN ENCOUNTER" which was syndicated by the Los Angeles Times for over six years. Together, Robert Snodgrass and Dawn teamed as MANTA Publications to produce marine and natural history posters and books. Dawn's poster projects are distributed to over 1,250 schools in Southern California, reaching over one million school children. You can see more of Dawn's illustrations at www.mantapublications.com.